# Natural

# Skincare

Sidra Shaukat qualified as a make-up artist from a leading international school of make-up, based in London. She has given many talks and demonstrations on this subject, and is an expert author, regularly appearing on the radio, as well as lending her expertise to magazines, newspapers, websites and ezines. She works as a freelance writer on health and beauty and lives in London.

# NATURAL SKINCARE

An indispensable guide to making your own natural skincare treatments

SIDRA SHAUKAT

ISBN 978-1-4452-1583-9

Notes from the author

Any information given in this book is not intended to be taken as a replacement for medical advice. Any person requiring medical attention should consult a qualified medical practitioner or suitable therapist.

I would like to dedicate this book to my immediate family: my late father, my beautiful mother, and my two amazing brothers.

## Acknowledgements

I am most grateful to Christine Wildwood for allowing me to use the basic moisturiser formula from her book Aromatherapy.
I would also like to thank my friends who have given me a few of their favourite treatments, for inclusion in this book.
Last, but not least, I would like to thank my family for their support and encouragement.

# Introduction

Growing up as a teenager in London, I was fascinated in the health and beauty rituals of my Asian heritage. My father would ask me to massage his head with my fingers and often add olive oil, which resulted in him keeping his thick, black, curly locks. My mother would massage my hair and scalp with oil, and comb them into two thick plaits, before I went to school. It wasn't cool but I think this resulted in me having a strong, healthy head of thick hair. I also loved the bridal rituals such as henna body painting on hands and feet, turmeric and yoghurt face masks, hair massaged with oil, natural sugaring hair removal of face and body (ouch!) with elaborate bridal hair and make-up.

There is growing consumer concern about chemicals in cosmetics & toiletries has resulted in increased demand for natural treatments.

Consumer concern about the possible harmful effects of parabens, sulphates and petrochemicals in personal care products is stimulating demand for products with natural ingredients. Natural products are now available in chemists, department stores, hair salons, beauty salons and spas.

Although many personal care products contain organic ingredients, certified organic products are also developing. The absence of official regulations for organic & natural personal care products is causing confusion for consumers who are unable to differentiate between legitimate natural products and conventional products with natural ingredients.

The best way to be absolutely sure you are using natural treatments is to follow the recipes in this book. You may want to invite your best friend over for some of the body treatments, and take it in turns to pamper each other. Most of the ingredients can be purchased at supermarkets or over the counter at larger chemists.

I spent years studied these tried and trusted recipes and came up with a few of my own. Enjoy making your own natural beauty treatments!

# My personal favourites

# 1. Face and body mask

When we think of a spa, the first thing we think of is the face and body covered with a green mud mask. Mud is fantastic for drawn out impurities from the skin to leave it glowing. A shower afterwards, as cold as you can bear, does wonders for the circulation.

I make a green clay and turmeric mask (turmeric is the yellow spice used in curries that is a great cleanser) by mixing one heaped tablespoon of green clay with 1 teaspoon of turmeric and mixing in live yoghurt or water if you prefer to make it a thick paste. If you have dry skin, you may want to add a little almond oil as required. This is enough for the face, make larger amounts for the body.

For oily skins, make a clay mask by mixing 1 whisked egg white with 1 tablespoon of fuller's earth (from chemists). Add a few drops of peppermint oil. Leave to dry and rinse with lukewarm water.

For dry skins, whisk 1 egg white, mix with 1 teaspoon of honey. Put on with a pastry brush and leave for 10 minutes. Wash off with cold water. The egg white gives this mask it's firming effect.

# 2. Baths

Your bath could be the equivalent of the spa bath for it's treatment properties. Cleopatra bathed in ass's milk to maintain her beauty, you can do the same too! Instead of ass's milk, you can add 1/4 pint of milk to the running water, or try 2 tablespoons of dried milk for the same effect. For added indulgence, a few drops of coconut essence will give the bath a tropical feel or try a few drops of coconut essence.

You can even create the effect of being in a thermal mud bath, as they do in Iceland, known for it's therapeutic effects, but just adding two or three tablespoons of green clay or fuller's earth to the running warm water. Soak for as long as you can for maximum benefit.

Sink into a warm bath scented with essential oils such as ylang-ylang, patchouli or rosemary(to clear the head) and lavendar, rose or camomile(to relax and soothe). Make sure the water is not too hot, as this overstimulates the nervous system and reduces the oils therapeutic powers. To make this indulgent soak extra special, put on some soft, soothing music, dim the lights in the bathroom and relax.

Bathing is an excellent way of relaxing after a hard day, but overusing commercial bath preparations can dry out your skin. Herbal baths are a refreshing alternative and are useful for cystitis sufferers who should avoid perfumed bubble baths and oils.

Tie a handful of fresh herbs in some muslin and place in the bath under the taps. Run the hot water first to release the scent from the herbs, and then top up with cold water. Try lovage, camomile (you could even use camomile teabags or herbal teabags!), mint or rosemary. Out of season, you can use dried herbs instead of fresh herbs. A drop of baby oil or wheatgerm oil or almond oil will stop your skin from drying out

# 3.Face and Body Scrubs and Exfoliators

There are various sea salt scrubs available on the market, but you can make your own. Raw sea salt is available from most supermarkets. Just mix with your favourite body oil, coconut oil is the best and rub vigorously all over the body. If you have sensitive skin, use granulated sugar instead of salt. Pay particular attention to the thighs to reduce and prevent cellulite.

Face scrubs will clear and brighten the complexion. Try a peach face scrub. Mash a large ripe peach, or liquidise using a food processor, mix a dessertspoon of oatmeal and rub all over the face. Rinse with warm water.

For a cheap face scrub and good exfoliator, mix

a small amount of sugar (preferably coarse such as granulated sugar) with your favourite cleanser. Massage over your face and wash off thoroughly. This acts in the same way as the expensive gels with exfoliating 'bead' in them, but is also gentler and much cheaper.

Try a nutty facial scrub for soft skin. Crush 4oz blanched almonds to a fine paste. Add 1 egg and 2 teaspoons of rosewater. Mix well. Using your fingertips, rub the mixture gently into the oily areas of your face and leave for 10 minutes before rinsing off well.

Grape juice is an effective natural exfoliator, removing dead skin cells and brightening the skin. Grapeseed oil is also a good, light moisturiser for normal to dry skin. Apples too are natural skin exfoliators, just cut a slice and whisk it over your face.

# 4. Hair removal

You may want to purchase wax and use it at home. I have created an extra special wax which is totally natural, made from just lemon juice, sugar, beeswax and honey. Or if you are feeling adventurous, try making your own.

In the Middle East, it is customary for the bride to be removed of all body hair. This is done using only natural products such as water, sugar, lemon juice and honey. One method is sugaring, which consists of purely water and sugar, heated gently, in different combinations, depending on the preference of the user.

Sugaring can be tricky to perfect but the time spent on practising can save time, money and the results are smooth, stubble-free skin which will remain hair-free for up to 8 weeks.

Try the 1 to 1 technique to start off with, using 1 tablespoon of water to 1 tablespoon of granulated sugar, in a metal pan, over gentle heat. when the sugar dissolves, and the liquid is the consistency of honey, allow it to cool slightly, and when at body temperature, spoon on mixture with a metal spoon or spatula.

Press mixture onto skin with hands or use a strip of cotton to press mixture against the skin. Pull off mixture with hands or take off strip against hair growth. Rinse with cold water.

You will notice a lovely sheen on the skin. You can vary combinations of sugar to water depending on how thick you like the mixture, and the consistency you find easiest to work with.

Once you are familiar with the technique, you can make larger amounts in the same proportions of sugar to water. You may find it messy at first, but after practice, it will be fuss and mess free!

## 5. Coconut oil - The perfect anti-ager for skin!

Coconut oil is rich in lauric acid, known for being antiviral, antibacterial and antifungal. This makes it a beauty tonic for both internal and external use. It will rejuvenate your skin and prevent wrinkles. It is an excellent massage oil. Used as a skin lotion for healthier, younger skin, it prevents destructive free-radical formation and protects against them. It helps to keep your skin from developing liver spots, and other blemishes caused by aging and over exposure to sunlight. It prevents sagging and wrinkling by keeping connective tissues strong and supple. In some cases it might even restore damaged or diseased skin.

The oil is absorbed into your skin and into the cell structure of the connective tissues, limiting the damage excessive sun exposure can cause.

Coconut oil will not only bring temporary relief to your skin, but it will aid in healing and repairing. It will have lasting benefits, unlike most lotions. It can help bring back a youthful appearance. The coconut oil will aid in removing the outer layer of dead skin cells, making your skin smoother. your skin will become more evenly textured with a healthy "shine". And the coconut oil will penetrate into the deeper layers of your skin and strengthen the underlying tissues.

Coconut oil gives a great sheen to your skin and condition it, which makes it the ideal skin anti-ager. Most anti-agers are made from chemicals, colours, and fragrance, but I have produced my own 100% skin anti-ager that comes in a clear, glass roll-on bottle. It may look hard and white when cool, but just holding it in your palm will make it liquid. Then just put it straight on your skin for the most conditioning, skin anti-ager ever. Used regularly, it will anti-ageing benefits, as it moisturises, conditions and shines all in one, preventing thinning and further lines and wrinkles of the skin.

# Natural
# Skincare

A mask for dry skins should not be used by someone with oily skin. If you have combination skin, however, use a treatment for oily skin only on the oily area - i.e., the T-zone of the face, which consists of forehead, nose and chin. Use the treatments for dry skin on the cheeks.

Anyone who has sensitive and fragile skin may want to avoid those treatments that contain acidic ingredients such as lemon and pineapple and possibly yoghurt. Exfoliation should be done gently, and use castor sugar instead of granulated sugar in the facial scrub containing sugar (see below).

# Exfoliating

Exfoliation, however it is executed, dislodges dead skin cells and speeds up the rate of cell renewal, and it is essential as you get older because your metabolism will slow down. Exfoliate by briskly brushing your skin with a loofah or massage mitt, which will get skin glowing. Alternatively, you can use finely ground oatmeal, a good gentle exfoliator, which makes an effective scrub before a shower. Simply rub it all over your skin.

Slough off dead skin cells with a mixture of 1 dessertspoon (10 ml) of finely ground almonds in a small carton of yoghurt. Massage over a clean complexion, then rinse away to leave your skin fresh and bright. If your skin is sensitive, use oatmeal instead of almonds.

Facial scrubs will clear and brighten the complexion. Try a peach scrub: mash or liquidize in a food processor a large ripe peach, mix in 1 dessertspoon (10 ml) of oatmeal and rub the mixture all over your face. Rinse with warm water.

An inexpensive facial scrub and good exfoliator can be made by mixing a small amount of sugar (preferably a coarse kind such as granulated sugar) with your favourite cleanser. Massage over your face and wash off thoroughly. This acts in the same way as the expensive gels with exfoliating 'beads' in them, but it is also gentler and much cheaper.

A nutty facial scrub can be made by crushing 4oz (100 g) of blanched almonds to a fine paste. Add an egg and 2 teaspoons (10 ml) of rosewater. Mix well. Use your fingertips to rub the mixture gently into the oily areas of your face. Leave it for 10 minutes before rinsing off well.

Grape juice is an effective natural exfoliator, removing dead skin cells and brightening the skin. Grapeseed oil is also a good, light moisturizer for normal to dry skin. Apples, too, are natural skin exfoliators - just cut a slice and whisk it over your face, using gentle, circular movements.

## Steaming

Steam unclogs pores to deep cleanse skin and soften it. Use a deep bowl filled with boiling water. Incline your face towards the steam and cover your head with a towel that will also go around the edges of the bowl to prevent steam escaping. The steam will penetrate and open the pores to bring the dirt to the surface. Tone with a non-alcoholic toner applied with cotton wool to close and tighten pores. Repeat weekly.

Tea acts as a mild tonic and conditioner, and a steaming tea facial will deep cleanse the skin. Pour hand hot tea into a bowl, add a few dashes of lemon rind and some marjoram and mint (dried herbs will do). Now lean your face over the bowl and cover your head with a towel. The heat and toning polyphenols in the tea will draw out impurities.

If you have dry skin, you may want to put some jojoba oil or evening primrose oil in the bowl to soften your skin if you are having a steaming session. Your skin will be able to absorb the oil as the pores open because the oil is in the form of tiny droplets.

# Face Masks

### Dry Skins

Apricot Oil Mask

Apricot kernel oil helps to maintain healthy skin and surface tissue. Because it is lightweight it soaks in quickly, so it is ideal for dry skins.

### Oily Skins

Lemon Mask

Lemon juice is a well-known cure for sallow, oily skin, and it is an ideal ingredient in face masks. Lemon juice is good for spot-prone skin because it neutralizes bacteria, and it contains high levels of vitamin C, which helps the skin to heal. Rub a slice of lemon over your face skin and rinse with cold water.

## All Skin Types

Exotic Fruit Mask

Fruits such as pineapple and papaya (or pawpaw) make good face masks because of the enzymes they contain, which help in the healing of wounds and are thus natural aids for the skin. Simply mash up the fruit, put it directly on your skin, leave for 10-15 minutes and rinse well to reveal a softer skin.

Kelp Mask

All skins benefit from a face mask that deep cleanses and refines without irritating. Take 3 tablespoons (60 ml) of powdered kelp (which is available from many herbalists), mix it with a little almond oil and rosewater (which is available from chemists). Apply to clean skin, leave for 10 minutes, rinse off and moisturize. Use weekly.

Strawberry Mask

Mashed strawberries make a stimulating face mask. Simply cover your face with the fruit puree for 5-10 minutes, then rinse well with tepid water. Apply a gentle toner if required.

# Problem Skins

Turmeric Mask

Many Asians believe that turmeric (a yellow spice available in powder form from most supermarkets) clears problem skins and promotes a smoother, softer complexion. In fact, in some parts of India, it is applied as a paste on both the bride and groom before the wedding to enhance their beauty!

You can mix 1 teaspoon (5 ml) of turmeric with a

few drops of warm water or milk and apply directly to the face. Alternatively, you can mix the turmeric with oatmeal (to exfoliate) and 1 teaspoon (5 ml) of yoghurt (which will act on bacteria, thus preventing spots). If you have dry skin you may wish to add a few drops of almond oil. Leave for 15-20 minutes or until it is dry.

Wash off with warm water and a mild soap or cleanser (try some of the recipes given), then use agentle, non-alcoholic toner (see page 28) with cotton wool to remove all traces of any remaining yellow stain. Use once a week, preferably at night, to enhance the skin's regenerating process, and either before or after a steam session to deep cleanse your skin.

## Skin Fresheners

A clean face flannel dipped in hot water placed over your face while you are lying down is an instant relaxer. To freshen or revive a tired face, dip a clean flannel in ice-cold water and place it over your face for 5 minutes. Your face will have a healthy, pink glow, signifying increased blood circulation to the surface of the skin. Your eyes will sparkle, too, and any puffiness will be eliminated.

Use cucumber to make several skin fresheners, which will be useful for normal skins, but especially for oily to combination skins. Puree (or grate if it is easier and quicker) a few slices of peeled cucumber with 2 tablespoons (40 ml) of natural yoghurt. Use this as a face mask for 30 minutes and rinse off with lukewarm water. Incidentally, yoghurt and cucumber dips are popular eaten after a hot curry to cool down! Use the leftovers of the dip to make a refreshing face mask!

## Treating Occasional Spots and Blemishes

To treat the odd blemish on all skin types dip a cotton bud in witch hazel and dab it directly on the area. At bedtime, soak a cotton wool ball in calamine lotion and apply it directly to the affected area. For stubborn spots, dot a tiny amount of a face pack such as clay, which dries hard, on the spot before going to bed. Toothpaste will also help to dry up spots. Put it directly on the spots with a brush and leave overnight.

Finally, give your skin an instant boost with a spray of mineral water. Fill a plant spray or a refillable perfume spray bottle for your handbag with bottled water and squirt it whenever your skin feels dull and sweaty. The spray also sets make-up, making it stay put all day!

## Moisturiser

Maximise the moisturising potential of any oil by soaking in the bath first before adding a few drops of your chosen oil to the water. This will ensure that the oil veils your skin and seals in some of the water absorbed in the soak.

Cleopatra is said to have bathed in ass's milk to enhance her beauty. Instead of ass's milk, try adding !4 pint (150 ml) of ordinary milk to the running water, or try 2 tablespoons (40 ml) of dried milk for the same effect. For extra indulgence and to give the bath a tropical feel, add a few drops of coconut essence.

## Looking After Your Legs

If you want to have silky smooth legs, exfoliate using a vegetable-based soap on a damp loofah or body brush, sweeping it up over wet skin in massaging movements. This will get rid of dead skin cells and loosen sebum blockages in pores. Legs have comparatively few sebaceous glands and are, therefore, prone to dryness. After a bath, slick on plenty of moisturizer, using upward massage movements. The steam from the bath will help to lock in moisture.

If you suffer from exceptionally flaky winter legs try this pack while you are in the bath. Mash an avocado, a banana and 1 tablespoon (20 ml) of thick cream and smooth it on to your skin. Leave for 10 minutes, then rinse off to reveal beautifully smooth legs.

Swollen ankles will be soothed by a solution of 2 tablespoon (40 ml) of Epsom salts in P/4 pints (1 litre) of lukewarm water. Bathe your feet and ankles for 10 minutes, then immerse them in cold water to reduce the swelling. Pat dry and massage gently until the aching stops.

## Scalp Massage

To release tension in your head, try scalp massage while you are in the bath. Start at the front of your head and press your finger tips firmly against your scalp. Gently move the scalp back and forth five times. Repeat, moving your fingers until you have covered your whole head.

Bathing is an excellent way of relaxing after a hard day, but overusing commercial bath preparations can dry out your skin.

# Natural Bath Preparations

Herbal baths are a refreshing alternative and are useful for sufferers from cystitis, who should avoid perfumed bubble baths and oils. Tie a handful of fresh herbs in a piece of muslin and fasten it under the bath taps. Run the hot water first to release the scent from the herbs and then top up with cold water. Try lovage, camomile (you could even use camomile or herbal teabags), mint or rosemary. If fresh herbs are not available, you can use dried herbs instead. A drop of baby, wheatgerm or almond oil in the water will stop your skin from drying out.

Sink into a warm bath scented with essential oils - ylang-ylang, patchouli or rosemary (to clear the head) or lavender, rose or camomile (to relax and soothe), for example. Make sure the water is not too hot, because this will overstimulate the nervous system and inhibit the therapeutic powers of the oils. To make this indulgent soak extra special, put on some soft, soothing music, dim the lights in the bathroom and relax.

Meditate in a warm, relaxing bath by closing your eyes, breathing in deeply and thinking beautiful thoughts - for example, you might like to imagine that you are relaxing on a favourite beach, listening to the sound of waves gently lapping on the shore and the breeze whispering in the palm trees overhead. Add a few drops of essential oil such as lavender, marjoram or juniper to increase the skin-soothing properties of your bath. Lavender has a deeply relaxing effect on body and mind.

The scent of orange is considered a pick-me-up and an antidepres-sant, so it is an ideal perfume to add to the bath at the end of a long, hard day. Add a drop of essential oil of orange or squeeze a seedless orange into the bathwater to refresh yourself and to aid relaxation.

## Body Scrubs

For an inexpensive body scrub that will effectively smooth away roughness and soften your skin, try sea salt. First, apply a little vegetable oil, such as almond or olive oil, to dampen your skin. Then give yourself a good rub with handfuls of sea salt, concentrating on any dry patches, such as your knees or elbows. Rinse off with a forceful shower of warm water followed by a cold shower to complete the toning treatment.

A bath containing a liberal sprinkling of sea salt is good for easing muscular aches and for curing minor skin problems.

Face packs need not be used only on your face.

Try them on other areas of your body such as your shoulders to unclog pores and slough off dead skin.

Clean greasy patches on shoulders and back with a soap-filled backbrush or loofah. If your back is badly blemished, use a deep-cleansing pack made from some fuller's earth mixed with a little water and lemon juice. Leave to dry and rinse off thoroughly with lukewarm water.

Facial scrubs and sponges to exfoliate the face can also be used on neglected areas such as rough knees and elbows. Massage thoroughly, then rinse. Apply a moisturizer or some handcream. (Handcream can be more economical than body lotion and just as effective.)

To treat badly neglected elbows, scrub them with a pumice stone or a bristle brush, then bleach them with lemon juice. Alternatively, try resting your elbows in two squeezed-out halves of a lemon. Smooth the skin afterwards by rubbing in plenty of moisturizer or your favourite oil.

# Skincare Highlights:

# Treatments for Specific Areas

# EYES

## Treatments for Puffy, Sore Eyes

Refresh your eyes by lying down, preferably in a quiet room, for 15 minutes with a slice of either raw potato or cucumber over each eye. Cotton wool pads soaked in witch hazel or in iced water can be used instead. You will notice that your eyes will sparkle and that any puffiness will have disappeared!

Tea has been used as a reviver for less than sparkling eyes for over two centuries! When you have had a late night or a long, hard day, relax with a couple of cooled teabags (the new round ones fit particularly well) over your eyelids for 10 minutes. You're guaranteed to feel instantly revived. It is not just the coolness of the teabag that makes you feel better, the polyphenols and tannin in tea have a mildly astringent and stimulating effect on the skin. Tea also causes skin to tighten slightly, and this helps to reduce puffiness and to remove dark circles.

## Natural Ways to Remove Eye Make-up

Cleanse away eye make-up and mascara with a trace of sweet almond oil and cotton wool. Wipe carefully from the nose to the outer eye. Leave a trace of oil on your skin to nourish it. Cotton buds (use one end for each eye) dipped into the oil and wiped gently over eye make-up and mascara will serve the same purpose. Rinse with a gentle toner, preferably one that is alcohol-, colour- and fragrance-free.

Baby oil is an effective eye make-up remover, especially for thick, waterproof mascara. Pour a small amount on to round cotton wool pads and wipe gently over the make-up, using one pad for each eye.

If your eyes are sensitive, dip cotton bud ends directly in the baby oil, remembering to use one end for each eye.

# Healthy Eyelashes and Eyebrows

Applying a little petroleum jelly to the root of your lashes each night will encourage healthy growth and make the lashes long and strong. For both day and evening make-up, petroleum jelly can be used to keep your eyebrows in shape and give them a good sheen. Work the jelly into your eyebrows with your fingers, but use only a tiny amount, because too much will look and feel greasy.

Petroleum jelly is also an effective, colourless mascara. Use an old eyelash comb to comb it through your lashes to make them look instantly thicker and longer. It will not clog, cause a sensitive reaction in the eyes nor, because it does not contain colour pigments as mascara does, will it create black marks all over your eyes when you cry!

Eyelash curlers obviate the need for mascara. However, if you have fair hair you might want to consider having your lashes dyed dark brown, which is far cheaper in the long run and is kinder to your eyes. However, do not attempt to dye your lashes yourself with hair dye. It is far better to seek professional help.

One way in which you can avoid expensive beauty treatments is by simply brushing your eyebrows into shape every night. Take an eyebrow brush (an old, soft toothbrush will do) and guide the eyebrows into the shape you want. Smooth a little petroleum jelly on to stubborn hairs to keep them in shape.

## HANDS

Your hands are on show to the world, so make sure that you try these simple but effective treatments for soft, supple hands.

Once a week, thoroughly moisturize your hands with an effective handcream and put on a pair of close-fitting, but not too tight cotton gloves. Leave them on for as long as possible, preferably overnight. The perspiration from your hands will blend with the cream to soften the skin thoroughly. This treatment will also help to reduce any prominent veins on the hands.

Keep your fingers slim and your hands soft by brushing the surface of hands and fingers with a cushioned brush - a hairbrush with cushioned balls on the spikes is ideal - towards the heart. This also improves circulation and helps to prevent unsightly raised veins.

# Hand Workout

Hands need exercise, too, to keep them soft and slim and to keep the joints functioning smoothly. Try the following easy exercises daily, and you will become more nimble fingered!
For a warm-up, form your hands into fist shapes, hold for a few seconds and then release. Repeat ten times.

Next join your hands together, pressing the palms flat against each other with your fingers straight as if you were praying. Fold over the fingers of one hand into the spaces between fingers of the other hand, making sure that they are folded over completely. Press your fingers as hard as possible for a few seconds and release. Repeat ten times. Finally, separate your hands and stretch out your fingers as far as possible; then relax. Repeat ten times. Rest and relax your hands and fingers for a few minutes.

Now you are ready to start the elastic band exercises. Put one end of a thick elastic band over the joint of the thumb, and the other end over the last joint of the first finger. Stretch the elastic band as far as it will go, relax and repeat ten times. Repeat the exercise separately with each of the second, third, and fourth fingers on both hands. Rest for a few minutes, then place the elastic band over the middle of thumb and all fingers to form a circle. Stretch the elastic band as far as possible and relax. Repeat ten times. These exercises are useful for keyboard users to prevent repetitive strain injury, and they will help those who suffer from rheumatism and arthritis.

## Dry Hands

Dry hands need intensive treatments. Mix 1 teaspoon (5 ml) of honey with a little vegetable, almond or olive oil and massage the lotion directly on to your hands, leaving it for as long as possible before rinsing with a soap-free cleanser. To exfoliate dead skin from the hands, blend some granulated sugar with a little oil (whichever type is handy) and rub well in. Rinse to reveal soft, new, gentle skin. This treatment also evens out the skin tone. Another way of achieving exfoliation is by moving a body brush loaded with a soap-free cleanser over your hands, including the palms. Rinse with cold water. This also improves circulation and speeds up skin renewal.

Help to make cracked, dry hands soft and supple by keeping a bottle of sweet almond oil next to the wash-basin. Add a few drops of oil to the warm water every time you wash your hands.

Once a week, try this hand treatment to pamper and soften your hands. Make a simple hand lotion by mixing 2 teaspoons (10 ml) of olive oil with 1 teaspoon (5 ml) of liquid honey. You can also add a drop of your favourite perfume or aromatherapy oil. Massage the mixture well into the hands, leave for a few minutes then wash off with warm, soapy water or a gentle shower gel.

Slough off dry, discoloured skin with a gentle scrub made from 1 tablespoon (20 ml) each of sunflower oil and salt mixed together well. Rub on to your skin and rinse with warm, soapy water or a gentle shower gel.

If you enjoy gardening but your hands suffer, smooth on this cream under your work gloves. Mix together 1 dessertspoon (10 ml) of almond oil and an equal amount of fuller's earth with two egg yolks. Smooth this on to your hands and put on your gloves. When you have finished gardening rinse away the lotion and your hands will have had a wonderful treatment while you have been working hard.

# Split Nails

Prevent nails from splitting and chipping by encouraging healthy nail growth with an oil bath. Gently warm some almond oil, then soak your fingertips in the oil for about 10 minutes. Take time to push cuticles back with a cotton bud after the soaking.

Protect your nails at all times from the damaging effects of gardening, housework, DIY and especially washing-up. Rubber gloves keep out the wet but wearing them too much actually softens the nails you are trying to strengthen. A good idea is to wear a pair of fine cotton gloves, which are available from chemists, under the rubber ones.

## Natural Manicure

For a quick home manicure, try the following. Clean off the old nail polish with plenty of remover. Shape your nails with an emery board, filing from side to centre in long, easy strokes. Give your hands a thorough wash, scrubbing your nails with lots of soapy water. Dry your hands carefully and work in lots of handcream, massaging the nail bed (see below). Then gently push back the cuticles with a cotton bud. Rinse away any traces of handcream and dry.

To encourage blood flow and thus growth, massage each nail bed for about a minute every evening with just a little sweet almond oil. The nail bed, where all the growth occurs, is just below the skin's surface immediately next to the cuticle.

Almond oil is an excellent nail and cuticle

conditioner. Simply work a little oil into each nail plate and leave for about 10 minutes before washing away. It will leave your nails looking and feeling wonderfully healthy. Alternatively, to encourage strong and healthy growth, rub petroleum jelly into your nails and nail bed at night. Wipe off any excess.

To achieve salon-effect French polish' nails, gently massage petroleum jelly into the nail and nail bed. Tissue off by gently rubbing with a soft tissue. For added shine, give a quick rub over with a nail buffer. This treatment is good for nails with or without nail polish.

For the natural look of a French manicure, try the following: make sure your nails are clean and neatly filed and that all old nail enamel has been removed. If your nails are discoloured, lighten them with a cotton bud dipped in white wine vinegar. Paint on a clear base coat, allow it to dry and put on a light coat of white pearl polish.

Leave to dry, then apply one coat of pale pink polish, finishing off with a clear top coat to add shine.

# LIPS

Petroleum jelly is essential for smooth, well-defined lips. It is the ideal colourless lip-gloss, which you can apply directly to your lips with your fingers or you can use a lip-brush or a cotton bud. Petroleum jelly can be used on its own or with lipstick. If you apply it before lipstick you will create a subtle sheen; applied on top of lipstick it will give a glamorous gloss. Applying petroleum jelly to your lips every night will condition them and help to avoid wrinkles.

## Cold Sores

Lemon juice is said to prevent cold sores. Citric acid quickly kills bacteria, making it a natural cleanser for grimy skin, and it would be well worth rubbing a slice of lemon over your lips if you get that itchy, pre-cold sore feeling.

## Natural Gloss

If one of your lips is darker than the other - the top lip is usually darker than the bottom lip - rub a slice of lemon over your top lip every day and rinse with cold water. This treatment also smoothes lips, so you may want to rub the lemon over both lips, but pay greater attention to the upper lip to make the colours even. Cocoa butter also has a smoothing effect on lips. Blend it in with your fingers, leaving it on overnight for maximum benefit.

If you want to give your lips a natural shade of red, cut a strawberry in half, using half for each lip. Rub the end of the strawberry over your lip and rinse with warm water. If you do not have any strawberries, crushed raspberries or redcurrants can also be used.

Smooth on moisturizer or night cream as a base either before make-up, to prevent lipstick leaving stains on your lips, or at night, to condition and soften, especially in cold weather.

## Chapped Lips

Honey is an excellent natural moisturizer. Simply rub it into your lips whenever they are sore or chapped. Leave it on for as long as possible.

## Lined Lips

If your lips have lines, try exfoliation to diminish them. Again, you can use sugar mixed with oil or scrubs, which should be rubbed on to your lips and rinsed off with warm water.

## Lip Exercises

Lips need exercise just like any other muscles. Simply stretch your lips by forming a wide smile and relax. Repeat five times. Curl your lips over your teeth and join your top and bottom teeth together. Relax. Repeat five times. Join your lips together and push forward as far as you can. Relax and repeat five times.

These exercises should help to prevent lines forming around your lips and make your lips fuller, thus preventing lipstick from 'feathering'.

# NECK

The skin on the neck is the first to show the signs of ageing, so pay especial attention here. The skin on the neck is very fine, and it is wise, therefore, to avoid heavy creams or moisturizers in this delicate area, because they could accentuate lines and stretch the skin, making the neck muscles sag even further.

Use only light oils on the neck; evening primrose oil is very good for the delicate skin on the neck, but any good oil will do the trick, even vegetable oil or olive oil. Sweet almond oil is inexpensive and effective. Massage in the oil of your choice, preferably at night so that the oil can penetrate into the skin for maximum benefit.

Always try to apply a moisturizer before you go out, because the skin in the neck area has very few sebaceous glands and it can get dry very easily. You will probably have noticed that you get hardly any blemishes on the neck, and this lack of sebaceous glands is the reason. If your neck looks too shiny after applying oil, gently tissue off the excess with a facial tissue.

Exfoliate gently everyday. You should, ideally, do this at night, when the skin repairs itself, to allow the skin to speed up its renewal process, dislodge dead cells and smooth out lines. This area responds to, and needs, exfoliation more than any other area - even more than the face. The emphasis is on being gentle. Harsh exfoliation in this delicate area can cause redness. Use a soft facial brush with a mild soap and warm water or massage with ground oatmeal.

The exercises for the neck, jaw and chin in the facial exercises section will improve the tone and condition of the skin and will help to smooth out any surplus lines.

Once a week, use a moisturizing mask. Try mixing 1 tablespoon (20 ml) of honey with 2 tablespoons (40 ml) of almond oil. Gently brush the lotion on to the neck with a soft brush (a pastry brush will do). Leave on for 30 minutes and rinse with warm water. Your skin will feel soft and smooth.

# Facial exercises

Most facial expressions do not use the muscles that keep contours youthful, and they can actually encourage wrinkles. Facial exercises, on the other hand, concentrate on specific, often lazy and unused muscles. You can build up facial muscles as you can those of your body, and, just as with other parts of your body, well-toned muscles mean firmer contours. Because the muscles are connected to the skin, you can keep the skin of your face well supported for longer.

## Relieving Tension

To soothe away tension and prevent premature wrinkles, try the following Chinese exercises.

Warm your hands by rubbing them briskly together. Place the middle fingers of each hand over your eyelids, just below the brow bone, and apply gentle pressure. Repeat three times.

Place the fingers of both hands horizontally and flat against your forehead. Press firmly against the bone, pulling the skin upwards and outwards very slightly; at the same time, move your eyebrows down and slowly close your eyes tight until you feel your eyes and your forehead pulling in opposite directions. Use your little fingers to pull between your brows to prevent frown lines. Relax and repeat three times.

To tone jaw muscles, which can accumulate

tension in your face, first locate the muscles by placing your fingers at the sides of your face and closing and opening your mouth a few times. Put your forefingers into the hollows at the hinges of your jaw. Pressing down, make smooth rotating movements with your fingertips for 30 seconds.

Rub the palms of your hands briskly up and down your cheeks. With your middle fingers, stroke from the centre of the forehead towards the temples, smoothing away frown lines. Using your fingertips and in decreasing circles, massage away tension. Use your eyes, nose, cheeks, jaw and mouth to pull as many faces as possible.

## Preventing Lines and Wrinkles

To prevent lines from forming around your nose and mouth, slightly curl your upper lip around your top teeth and your lower lip around your bottom teeth. Press your lips firmly together and blow as hard as you can, keeping your mouth closed, for a count of six. Relax and repeat three times.

Form your mouth into an O shape, pushing your jaw and upper lip forwards. Slightly curl your upper lip over your top teeth, hold for a count of six, then relax. Repeat the exercise but with your head tilted back. Then turn your head to the right, and repeat the O movement. Do exactly the same with your head turned to the left. Repeat the entire sequence three times.

To tone up your cheek muscles, fill your mouth with water - it's pressure that does the work - and hold it there for as long as possible. Then relax.

For strong cheek and chin muscles, smile with your mouth closed, raising your cheeks and the corners of your lips up towards your eyes. Hold for a count of six, then, in the same position, jut your jaw forwards and hold this position for a count of six. Relax, then grin, this time with your mouth slightly open and stretching the corners of your mouth out to the sides. Count to six, then, in the same position, jut your jaw forwards and count to six. Relax. Repeat the entire exercise three times.

To help prevent a double chin from developing, jut out your chin and, using the back of the fingers of each hand, gently tap upwards for a stimulating massage. Use a firm, bouncy motion, drumming, but not slapping too hard, against the skin with each hand alternately for 30 seconds.

To strengthen your neck muscles, purse your lips together and place your forefingers on your neck, one each side of your windpipe. Massage smoothly, up and down, with firm, gentle strokes, or 30 seconds.

# Exercises for Your Eyes

Prevent 'crow's feet' from developing by pressing the middle and third fingers horizontally at the corner of your eyes. Close your eyes slowly, using the muscles under the eyes. Count to six and repeat three times.

The formation of droopy skin above the eyes can be prevented by placing your index fingers under your eyebrows. Press on the bone and lift upwards slightly, at the same time closing your lids tightly. When you feel a slight pull on the skin, count to six. Relax, open your eyes and repeat three times.

To avoid eye fatigue, press firmly for a second at the pressure point on the bridge of your nose. Repeat five times.

Ease eyestrain by looking up, then down, then to the far right and the far left. Look from top right diagonally to bottom left, and from top left to bottom right. Look up, then circle your eyes from the side in a clockwise direction. Then circle in an anti-clockwise direction. Hold your thumb about 12 in (30 cm) from your eyes, focus on imaginary walls behind and in front of your thumb.